Masteri

SURVEY

M

An original Book by

Jim Crume P.L.S., M.S., CFedS

Co-Authors

Cindy Crume

Bridget Crume

Troy Ray R.L.S.

Mark Sandwick P.L.S.

Mark Lull

KINDLE - PRINTED EDITIONS

*****

PUBLISHED BY:

Jim Crume P.L.S., M.S., CFedS

Mastering the RPN & ALG Calculator

First publication: November 2015

Cover photo: HP 35s & TI-30Xa

TERMS AND CONDITIONS

TABLE OF CONTENTS

INTRODUCTION

Straight forward Step-by-Step instructions.

This book is just one part in a series of digital and paperback books on Surveying Mathematics Made Simple. The subject matter in this book will utilize the methods and formulas that are covered in the books that precede it. If you have not read the preceding books, you are encouraged to review a copy before proceeding forward with this book.

For a list of books in this series, please visit:

http://www.cc4w.net/ebooks.html

Prerequisites for this book:

A knowledge of geometry, algebra, trigonometry and Coordinate Geometry are required for the explanations shown in this book.

The following books of this series are recommended to complete some of the step by step processes in this book:

Bearings and Azimuths - Book 1
Create Rectangular Coordinates - Book 2
Inverse Between Rectangular Coordinates - Book 3
Circular Curves - Book 4
Parcel Boundaries - Book 5
Spiral Curves - Book 6
The Myth About Spiral Curve Offsets - Book 7
Intersections - Book 8
Coordinate Transformation - Book 9
Vertical Curves - Book 10

These books contain formulas, step by step solutions and examples.

Throughout this book, tips will be given to help explain or give directions on the subject matter.

DEFINITIONS

HP: Hewlett-Packard calculators. Through the years, HP has released several calculators that varied in their mathematical, programmability, and I/O capabilities.

TI: Texas Instrument calculators. Through the years, TI has released several models of calculators that vary in their mathematical, programmability, and I/O capabilities.

RPN: Reverse Polish Notation, also known as postfix notation, all operations are entered after the operands on which the operation is performed. This type of calculator usually cost more than the ALG type.

ALG: Algebraic Notation, with this mode the precedence of mathematical operators is taken into account. Texas Instrument pioneered algebraic notation with precedence and parentheses in 1974. This type of calculator usually cost less than the RPN type.

BACKGROUND INFORMATION

I have taken many exams over the years and continue to take them as I advance my career, license renewals and on-going professional development requirements.

The exams that I have taken span several states as a Professional Land Surveyor, with the Bureau of Land Management, for the Mineral Surveyor appointment and as a Certified Federal Surveyor. I have taken many other exams that are non-surveying related. Some of which required the use of a calculator.

Most of the exams I have taken have one thing in common, a calculator, which is needed to solve complex math problems.

Programmable calculators were not allowed in the testing facility when I started taking exams. I learned early on not to rely on programs that are entered into calculators to take the exams. I decided to master the calculator, learn the surveying math equations and the steps to solve them.

My choice of calculator is the Hewlett Packard (HP). I started out with the HP 25 many years ago. It had all the functions I needed to solve most math equations.

My current calculator is a HP 35s and a HP 15C app on my iPhone. I use my iPhone app most of the time because it is handy and always with me. I have, and can use, an algebraic calculator when needed. I will cover both types of calculator in this book.

I solve equations manually every now and then. It helps me to stay sharp. I use computer programs most of the time in order to get a project done as quickly as possible, to meet the needs of the client and to stay within budget. I really like being on the cutting edge of technology. If only we could use this technology when taking an exam. The reality is, that is never going to happen.

The most popular option is to select a calculator, from the approved list, that has programming capabilities to enter coding that will be needed to solve various surveying math problems. This means that you will need to learn the program coding and routines so that you use the correct one to solve a problem (i.e. bearing-bearing intersection). You are going to need to spend a considerable amount of time learning the program and its functions in order to be fast enough with it to take a timed test. There is a risk with this method of preparing for the exam. If the calculator's battery dies during the exam, then you will lose the program which you worked so hard to master. You won't have time during the exam to reprogram the calculator nor will you be able to remember all of the coding. You will need to have a lot of faith in the programming code and the results the program is calculating, especially if you got it from a third party. That is a lot of risk to add to a career building exam.

The **BEST** option is to learn the location of the keys and functions already built into the calculator. In my experience, you will be much better prepared to take a timed exam by mastering your calculator. I have expressed this many times to my staff and others, that mastering the calculator is the **KEY** to passing a survey exam. There are no shortcuts or calculator programs that will replace the knowledge needed to pass the exam. Remember that you are the master of your destiny.

Once you have passed the test, you can focus on mastering the computer and calculator programs which will greatly help you in your surveying career.

NCEES CALCULATOR POLICY

The calculators that have been approved for the NCEES exams at the time of the writing of this book are as follows:

Casio

fx-115 MS
fx-115 MS Plus
fx-115 MS SR
fx-115 ES
fx-115 ES Plus

Hewlett Packard

HP 33s *
HP 35s * *(Covered in this book)*

Texas Instruments

TI-30Xa * *(Covered in this book)*
TI-30Xa SOLAR *
TI-30Xa SE *
TI-30XS Multiview
TI-30X IIB
TI-30X IIS
TI-36X II
TI-36X SOLAR
TI-36X Pro

You will need to check the NCEES.ORG website for the official list of approved calculators since it does change from time to time.

* Calculators with DMS <> D.ddd conversion function.

TYPES OF CALCULATORS

There are two types of scientific calculators, Reverse Polish Notation (RPN) and Algebraic Notation (ALG). I have used both over the years. I prefer RPN because it takes fewer keystrokes to solve a problem. It is a little more difficult to learn but once you do, it is a real time saver. The ALG calculator is the most popular though.

This book will cover the HP 35s (RPN) and the TI-30Xa (ALG).

It can be intimidating when you first look at these calculators. You immediately think that you need to understand all of the functions that are shown in order to use the calculator efficiently. That is a MYTH. There are only a handful of functions that you need to master in order to solve the equations on a surveying exam. Out of the 120+ available functions, you only need to use about 45 of them.

HP 35s

Lets take a look at the HP 35s. **Figure 1** is the standard keyboard for the HP 35s. There are about 123 functions shown on the standard keyboard. "You will, over time, use" several of these functions. For the purposes of this book, we are going to only focus on the minimum functions needed to solve equations that you will likely encounter during a surveying exam.

Figure 2 shows only the functions that will be covered in this book and that you will need to master. It is less intimidating when you strip away the unneeded functions.

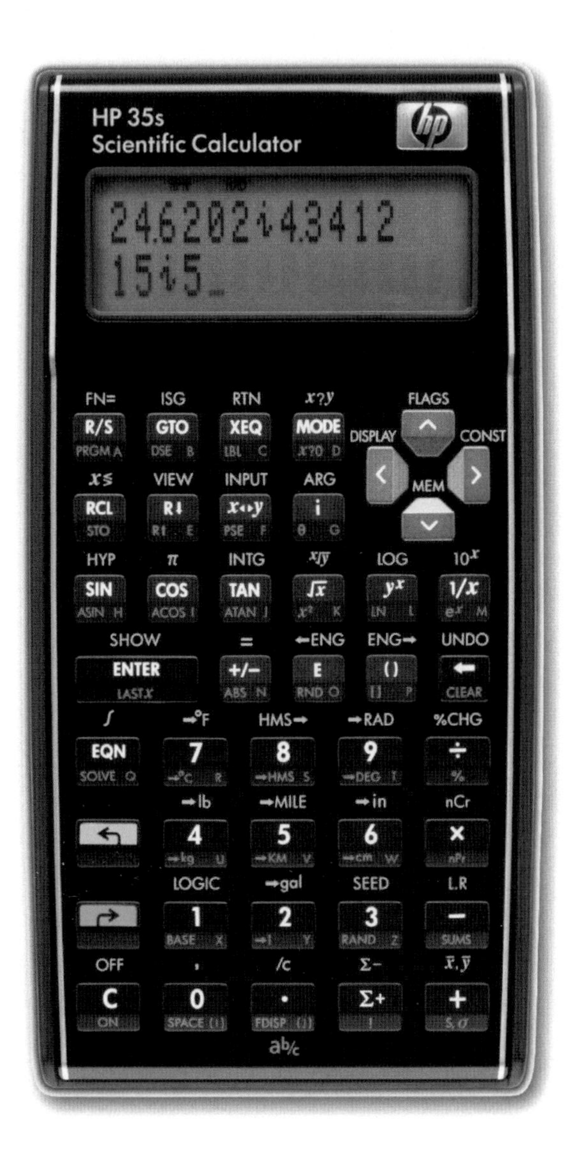
HP 35s
Scientific Calculator
hp
24.6202i4.3412
15i5_
FN= ISG RTN x?y FLAGS
R/S GTO XEQ MODE
DISPLAY MEM CONST
PRGM A DSE B LBL C x?0 D
x≶ VIEW INPUT ARG
RCL R↓ x↔y i
STO R↑ E PSE F θ G
HYP π INTG LOG 10^x
SIN COS TAN √x y^x 1/x
ASIN H ACOS I ATAN J x² K LN L e^x M
SHOW = ←ENG ENG→ UNDO
ENTER +/− E ( ) ←
LASTx ABS N RND O [ ] P CLEAR
∫ →°F HMS→ →RAD %CHG
EQN 7 8 9 ÷
SOLVE Q →°C R →HMS S →DEG T %
→lb →MILE →in nCr
4 5 6 ×
→kg U →KM V →cm W nPr
LOGIC →gal SEED L.R
1 2 3 −
BASE X →l Y RAND Z SUMS
OFF /c Σ− x̄,ȳ
C 0 . Σ+ +
ON SPACE FDISP ! s,σ
ab/c

Figure 1.

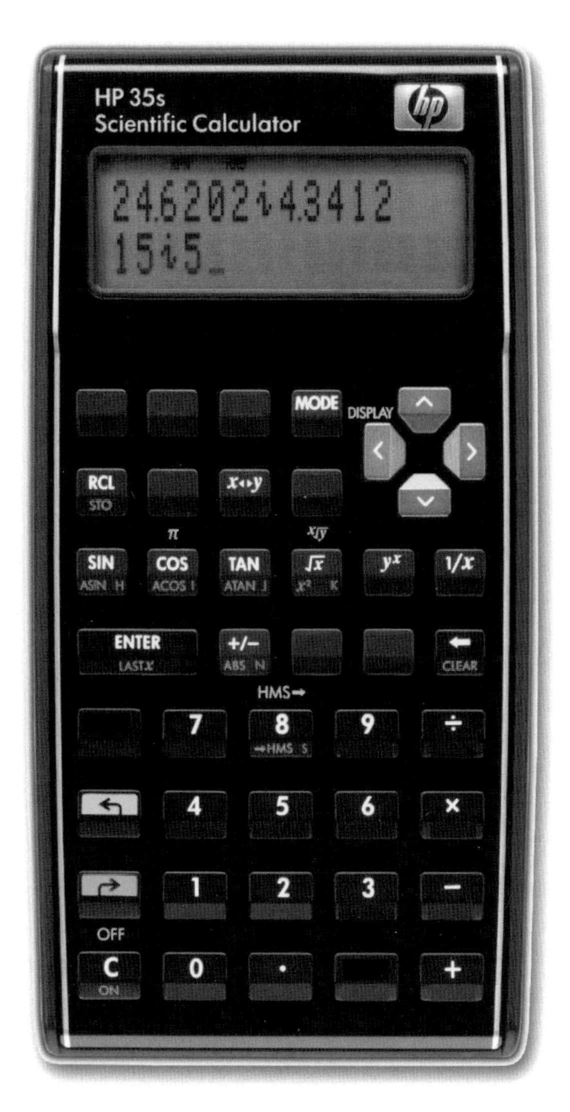
HP 35s
Scientific Calculator
24.6202i4.3412
15i5_
MODE
DISPLAY
RCL
STO
x↔y
π
x√y
SIN
ASIN H
COS
ACOS I
TAN
ATAN J
√x
x^2 K
y^x
1/x
ENTER
LASTx
+/−
ABS N
CLEAR
HMS→
7
8
→HMS S
9
÷
4
5
6
×
1
2
3
−
OFF
C
ON
0
•
+

Figure 2.

(HP) Minimum Functions Needed	
DISPLAY <	Sets the display format: (1 **FIX**, 2 SCI, 3 ENG, 4 ALL) Press [↰] [Display] then type [1] for <u>FIX</u> then type a number between [0] to [9] to set number of decimal places.
MODE	Sets the calculator Mode: (1 **DEG**, 2 RAD, 3 GRD, 4 ALG, 5 **RPN**) Press [MODE] then type [1] for <u>DEG</u> mode. Press [MODE] then type [5] for <u>RPN</u> mode. The screen should look like the following: RPN Y 0.707106781 X 1.414213562
RCL STO	Press [↱] [STO] then a number between [0] and [9] to store X. Press [RCL] then a number between [0] and [9] to recall the stored value to X.
x↔y	Press [x<>y] to exchange the values of X and Y.
SIN ASIN H	Press [SIN] to calculate the Sin of X. Press [↱] [ASIN] to calculate the Inverse Sin of X.

π COS ACOS	Press [←] [π] to recall PI to X. Press [COS] to calculate the Cosine of X. Press [→] [ACOS] to calculate the Inverse Cosine of X.
TAN ATAN	Press [TAN] to calculate the Tangent of X. Press [→] [ATAN] to calculate the Inverse Tangent of X.
$\sqrt[x]{y}$ $\sqrt{x}$ x^2	Press [←] [$\sqrt[x]{y}$] to calculate the X root of Y. Press [$\sqrt{x}$] to calculate the square root of X. Press [→] [x^2] to square X.
y^x	Press [y^x] to calculate the X power of Y.
1/x	Press [1/x] to calculate the inverse of X.
ENTER LASTx	Press [→] [LASTx] to recall the last X value. Press [Enter] to insert a numeric number into X.
+/− ABS	Press [→] [ABS] for the absolute value of X. Press [+/-] to change the sign of X.
← CLEAR	Press [→] [Clear] then select (1 X, 2 VARS, 3 ALL) to clear the stack. Press [←] to clear the last digit entered.

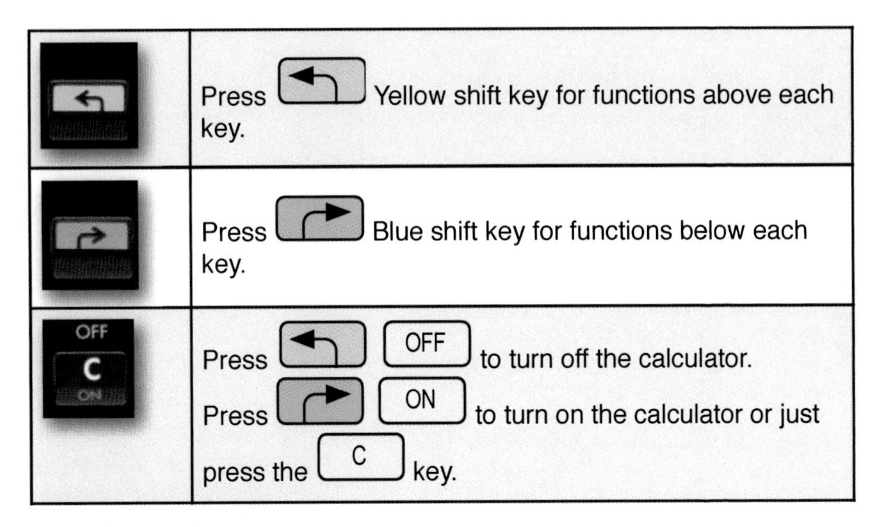

	Press [↰] Yellow shift key for functions above each key.
	Press [↱] Blue shift key for functions below each key.
OFF C ON	Press [↰] [OFF] to turn off the calculator. Press [↱] [ON] to turn on the calculator or just press the [C] key.

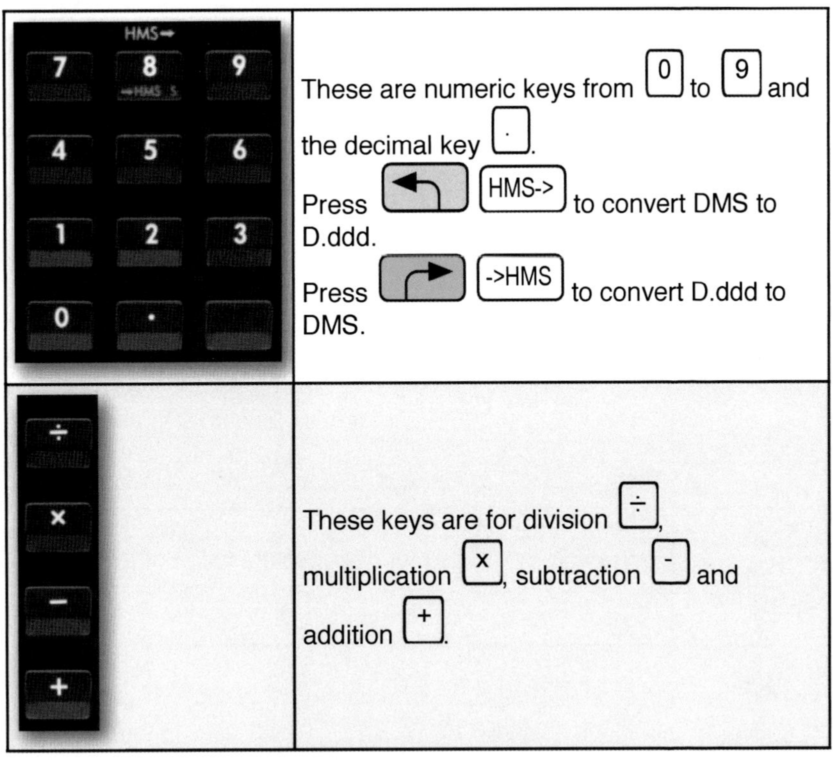

HMS→ 7 8 9 →HMS 4 5 6 1 2 3 0 .	These are numeric keys from [0] to [9] and the decimal key [.]. Press [↰] [HMS->] to convert DMS to D.ddd. Press [↱] [->HMS] to convert D.ddd to DMS.
÷ × − +	These keys are for division [÷], multiplication [x], subtraction [-] and addition [+].

HP Fundamentals

The first thing you want to do is to set the Mode of the calculator to the notation and angular units that you want to work in.

Press [MODE] [5] to set the calculator to **RPN** for the HP examples in this book.

Press [MODE] [1] to set the angular mode to **DEG** for the HP examples in this book.

The Modes that are available are (1 DEG, 2 RAD, 3 GRD).

If you are working in GRAD (a.k.a. GON) then select 3 for GRD mode.

Press [↰] [Display] [1] then a number between [0] to [9].

This will set the number of decimal places that are displayed.

RPN
0.707106781
1.414213562

The Stack

The screen will always display the values of X and Y of the stack.

RPN
Y 0.707106781
X 1.414213562

There are also two hidden rows for the Z and T values just above the X and Y rows. You won't need to see these values you just need to know that they are there.

The value in the X row is the one that is always in focus and all calculator functions will use this value when performing any functions. Depending on the function being performed, the value in the Y row will be used in conjunction with the X row such as adding, subtracting, etc.

As numbers are entered, added, subtracted and so forth, the values in each of the rows will move up and down the stack automatically.

Just remember the X row is always in focus.

TI-30Xa

Lets take a look at the TI-30Xa. **Figure 3** is the standard keyboard for the TI—30Xa. There are about 76 functions shown on the standard keyboard. You will over time use several of these functions. For the purposes of this book, we are going to only focus on the minimum functions needed to solve equations that you will likely encounter during an exam.

Figure 4 shows only the functions that will be covered in this book and that you will need to master. It is less intimidating when you strip away the unneeded functions.

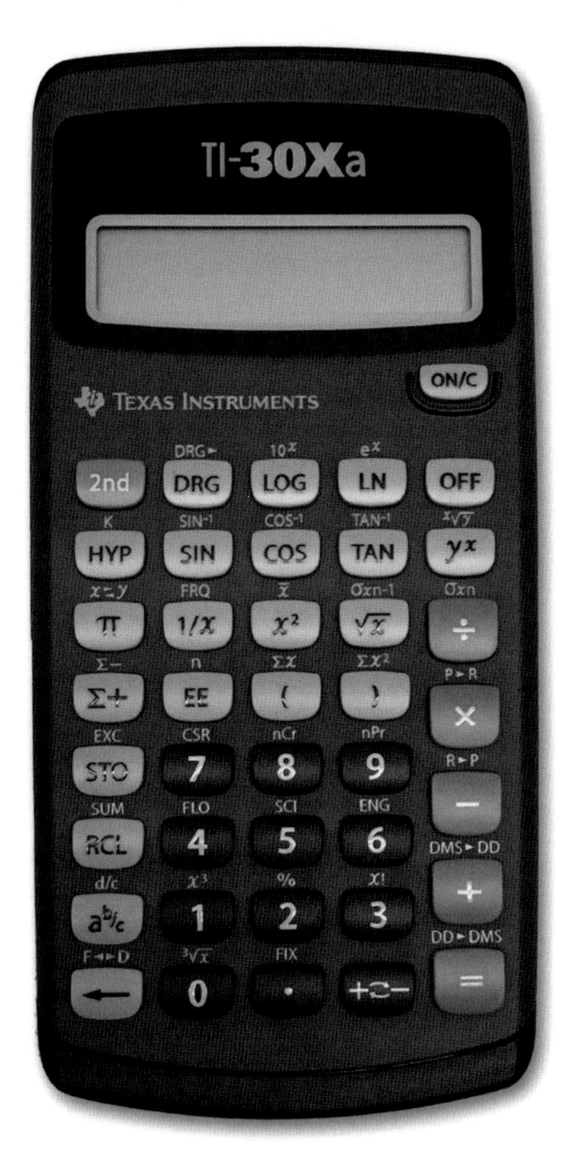
TI-30Xa
TEXAS INSTRUMENTS
ON/C
DRG► 10x ex
2nd DRG LOG LN OFF
K SIN-1 COS-1 TAN-1
HYP SIN COS TAN yx
FRQ σxn-1 σxn
π 1/x x2 √x ÷
Σ- n Σx Σx2 P►R
Σ+ EE ( ) ×
EXC CSR nCr nPr R►P
STO 7 8 9 −
SUM FLO SCI ENG DMS►DD
RCL 4 5 6 +
d/c x3 % x! DD►DMS
1 2 3 =
F◄►D FIX
0 .

Figure 3.

TI-30Xa
TEXAS INSTRUMENTS
ON/C
DRG▸
2nd
DRG
OFF
SIN⁻¹
COS⁻¹
TAN⁻¹
x√y
SIN
COS
TAN
yx
x⇆y
π
1/x
x²
√x
÷
×
STO
7
8
9
FLO
−
RCL
4
5
6
DMS▸DD
x³
+
1
2
3
DD▸DMS
³√x
FIX
0
.
+⇄−
=

Figure 4.

(TI) Minimum Functions Needed	
ON/C	Press ON/C turns the calculator ON and/or clears the screen
2nd	Press 2nd Green shift key for functions above each key.
OFF	Press OFF turns the calculator OFF.
DRG► DRG	Press 2nd DRG> converts angle-unit setting between degrees, radians and GRAD. Press DRG cycles angle-unit setting between degrees, radians and GRAD.
SIN^{-1} SIN	Press 2nd SIN^{-1} to calculate the Inverse Sin. Press SIN to calculate the Sin.
COS^{-1} COS	Press 2nd COS^{-1} to calculate the Inverse Cos. Press COS to calculate the Cos.
TAN^{-1} TAN	Press 2nd TAN^{-1} to calculate the Inverse Tan. Press TAN to calculate the Tan.
$\sqrt[x]{y}$ y^x	Press 2nd $\sqrt[x]{y}$ to calculate the X root of Y. Press y^x to calculate the X power of Y.

x<>y / π	Press [2nd] [x<>y] to exchange the values of X and Y. Press [π] to recall PI.
1/x	Press [1/x] to calculate the inverse of X.
x^2	Press [x^2] to square X.
$\sqrt{x}$	Press [$\sqrt{x}$] to calculate the square root of X.
STO	Press [STO] then a number between [1] and [3] to store X.
RCL	Press [RCL] then a number between [1] and [3] to recall the stored value to X.
←	Press [←] to clear the last digit entered.
÷	Press [÷] for division.
×	Press [x] for multiplication.
−	Press [-] for subtraction.

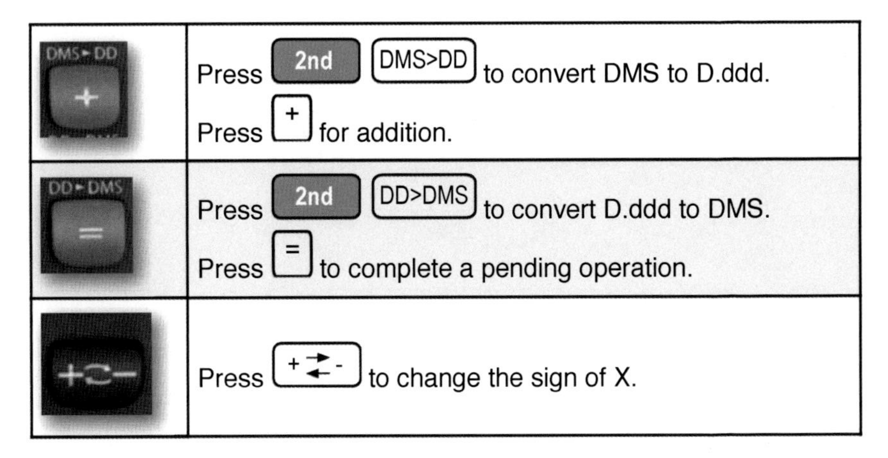

DMS►DD +	Press 2nd DMS>DD to convert DMS to D.ddd. Press + for addition.
DD►DMS =	Press 2nd DD>DMS to convert D.ddd to DMS. Press = to complete a pending operation.
+⇄−	Press +⇄- to change the sign of X.

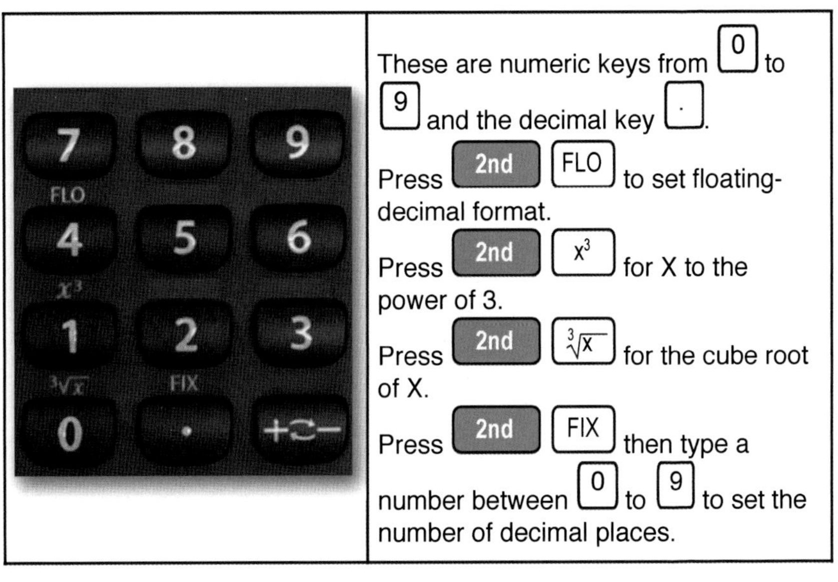

7 8 9 FLO 4 5 6 x^3 1 2 3 $\sqrt[3]{x}$ FIX 0 . +⇄−	These are numeric keys from 0 to 9 and the decimal key . Press 2nd FLO to set floating-decimal format. Press 2nd x^3 for X to the power of 3. Press 2nd $\sqrt[3]{x}$ for the cube root of X. Press 2nd FIX then type a number between 0 to 9 to set the number of decimal places.

TI Fundamentals

The first thing you want to do is to set the Mode of the calculator to the notation and angular units that you want to work in.

Press DRG until DEG is displayed. DEG mode will be used for the TI examples in this book.

The Modes that are available are (DEG, RAD, GRAD).

If you are working in GRAD (a.k.a. GON) then select GRAD mode.

Press [2nd] [FIX] then a number between [0] to [9].

This will set the number of decimal places that are displayed.

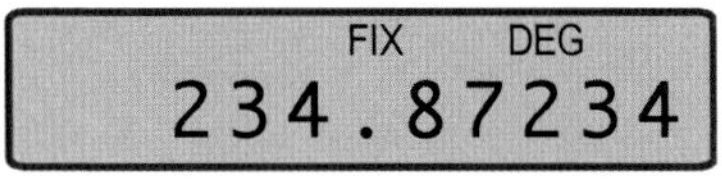

FIX DEG
234.87234

The Stack

The screen will display only the value of X.

FIX DEG
234.87234

There is a hidden row for the Y value. You won't need to see this value, you just need to know that it is there.

The value in the X row is the one that is always in focus and all calculator functions will use this value when performing their function. Depending on the function being performed, the value in the Y row will be used in conjunction with the X row such as adding, subtracting, etc.

As numbers are entered, added, subtracted and so forth, the values in each of the rows will move up and down the stack automatically.

Just remember the X row is always in focus.

DEGREES CONVERSION

One of the **KEY** elements in performing calculations is to convert **Degrees-Minutes-Seconds** to **Decimal Degrees** before applying any trigonometric functions. Most examinees forget to do this step which results in the wrong answers. They then forget to convert **Decimal Degrees** back to **Degrees-Minutes-Seconds** for the final answer.

The two calculators represented in this book both have a built in function to convert from (DMS to D.ddd) and (D.ddd to DMS). You really need these functions for taking an exam. Without them you will need to rely on base 60 conversion calculations which can be intense during an exam.

See "Bearings and Azimuths - Book 1" for more information base 60 conversion.

(HP 35s) DMS to D.ddd	
Given DMS	45°15'30" Convert to D.ddd
Keystrokes	45.1530 [↰] [HMS->] to convert to D.ddd
Result D.ddd	RPN Y 0.00000 X 45.25833 D.ddd

(TI-30Xa) DMS to D.ddd	
Given DMS	45°15'30" Convert to D.ddd
Keystrokes	45.1530 [2nd] [DMS>DD] to convert to D.ddd

Result D.ddd	FIX DEG 45.25833 D.ddd

(HP 35s) D.ddd to DMS	
Given D.ddd	33.78923° Convert to DMS
Keystrokes	33.78923 [↱] [->HMS] to convert to DMS
Result DMS	RPN Y 0.00000 X 33.47212 33°47'21.2"

(TI-30Xa) D.ddd to DMS	
Given D.ddd	33.78923° Convert to DMS
Keystrokes	33.78923 [2nd] [DD>DMS] to convert to DMS
Result DMS	FIX DEG 33°47'21"2 33°47'21.2"

This exercise is to help ingrain the process to muscle memory.

Repeat the above process for the following values:

Exercise 1 (DMS to D.ddd)		
DMS	**HP 35s**	**TI-30Xa**
78°33'12"	???	???

123°12’56”	???	???
89°39’54”	???	???

Exercise 2 (D.ddd to DMS)		
D.ddd	**HP 35s**	**TI-30Xa**
92.34521°	???	???
98.89765°	???	???
129.65754°	???	???

The answers can be found at the end of the book.

GRAD (a.k.a. GON)

GRAD (a.k.a GON) are easier to work with than Degrees-Minutes-Seconds. The reason being is that there is no conversion needed. GRAD are always in decimal format.

To use the trigonometric functions in the calculators, they need to be set to GRAD mode.

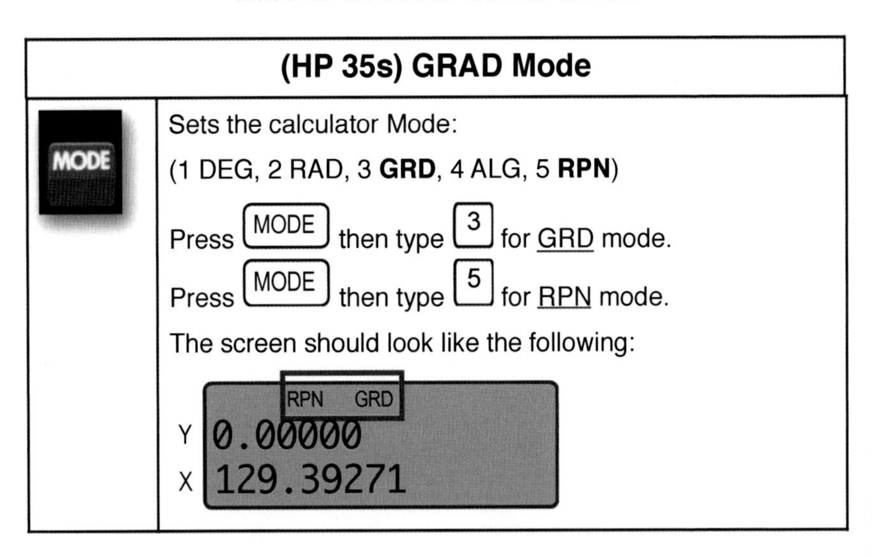

(HP 35s) GRAD Mode

Sets the calculator Mode:

(1 DEG, 2 RAD, 3 **GRD**, 4 ALG, 5 **RPN**)

Press MODE then type 3 for GRD mode.

Press MODE then type 5 for RPN mode.

The screen should look like the following:

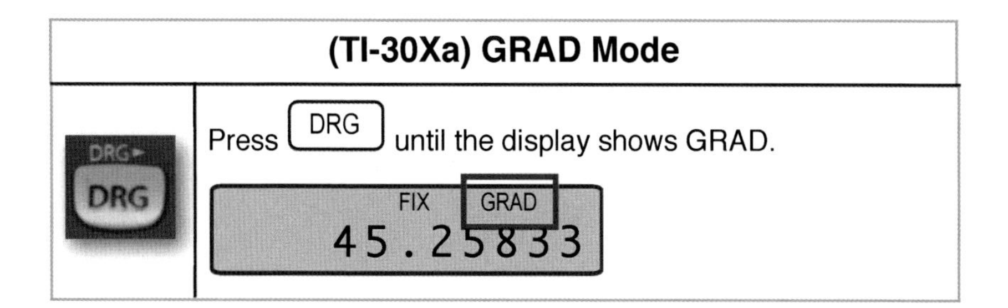

CONVERSIONS (D.ddd vs GRAD)

(HP 35s) D.ddd to GRAD	
Given D.ddd	234.87234°
Keystrokes	234.87234 [Enter] 400 [x] 360 [÷] **NOTE**: The 400 (GRAD) and 360° (DEG) are constants for a full circle.
Result	RPN Y 0.00000 X 260.96927 in GRAD **NOTE**: Calculator stays in DEG mode. You cannot use GRAD for trigonometric functions while in DEG mode.

(TI-30Xa) D.ddd to GRAD	
Given D.ddd	234.87234°
Keystrokes	234.87234 [x] 400 [÷] 360 [=] **NOTE**: The 400 (GRAD) and 360° (DEG) are constants for a full circle.

Result	FIX DEG 260.96927 in GRAD NOTE: Calculator stays in DEG mode. You cannot use GRAD for trigonometric functions while in DEG mode.

(HP 35s) GRAD to D.ddd	
Given GRAD	260.96927
Keystrokes	260.96927 [Enter] 360 [x] 400 [÷] **NOTE**: The 400 (GRAD) and 360° (DEG) are constants for a full circle.
Result	RPN Y 0.00000 X 234.87234 in DEG D.ddd **NOTE**: Calculator stays in DEG mode.

(TI-30Xa) GRAD to D.ddd	
Given GRAD	260.96927
Keystrokes	260.96927 [x] 360 [÷] 400 [=] **NOTE**: The 400 (GRAD) and 360° (DEG) are constants for a full circle.
Result	FIX DEG 234.87234 in DEG D.ddd NOTE: Calculator stays in DEG mode.

Keeping track of which mode the calculator is in is very important. It is better to pick a mode and stay in it than to switch back and forth.

Since both calculators will work in GRAD mode, there is no need to work in RADIANS therefore that will not be covered in this book.

TRIGONOMETRIC FUNCTIONS

You will be required to calculate SIN, COS and TAN functions as well as the inverse of these functions throughout the exam. In this section we will combine the DMS to D.ddd conversion and trigonometric functions.

(HP 35s) SIN, COS or TAN	
Given DMS	34°44'56"
Keystrokes	34.4456 [↰] [HMS->] [SIN]
Result	RPN Y 0.00000 X 0.56998 SIN

(TI-30Xa) SIN, COS or TAN	
Given DMS	23°55'23"
Keystrokes	23.5523 [2nd] [DMS>DD] [COS]
Result	FIX DEG 0.91409 COS

(HP 35s) ASIN, ACOS or ATAN	
Given	0.44362
Keystrokes	0.44362 [↱] [ATAN] [↱] [->HMS]
Result	RPN Y 0.00000 X 23.55229 23°55'22.9"

(TI-30Xa) SIN-1, COS-1 or TAN-1	
Given	0.44362
Keystrokes	0.44362 [2nd] [TAN-1] [2nd] [DD>DMS]
Result	FIX DEG 23°55'22"9 — 23°55'22.9"

(HP 35s) DMS to GRAD to SIN	
Given DMS	34°44'56"
Keystrokes	34.4456 [↰] [HMS->] 400 [x] 360 [÷] [SIN]
Result	RPN GRD Y 0.00000 X 0.56998 SIN **NOTE**: Calculator must be in GRD mode.

The last operation was shown just to demonstrate the keystrokes to go from DMS to GRAD then take the SIN. The calculator must be in GRD mode for this to work properly. It is recommended not to mix Degrees and GRAD. Pick the angular unit you work in most of the time and stick with it.

WORKING WITH EQUATIONS

Now that the basics have been established, it is time to evaluate an equation and properly enter it into the calculator.

Bearings and Azimuths

Equation 1: Bearing = N 20°45'10" E + 34°20'30"

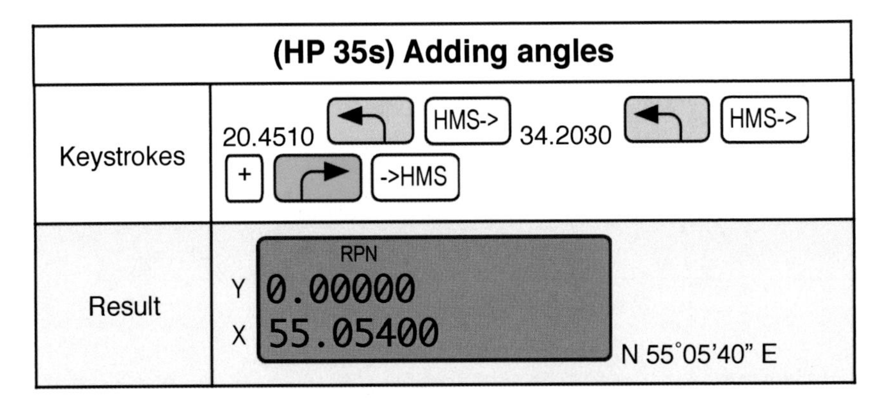

(HP 35s) Adding angles	
Keystrokes	20.4510 [↰] [HMS->] 34.2030 [↰] [HMS->] [+] [↱] [->HMS]
Result	RPN Y 0.00000 X 55.05400 N 55°05'40" E

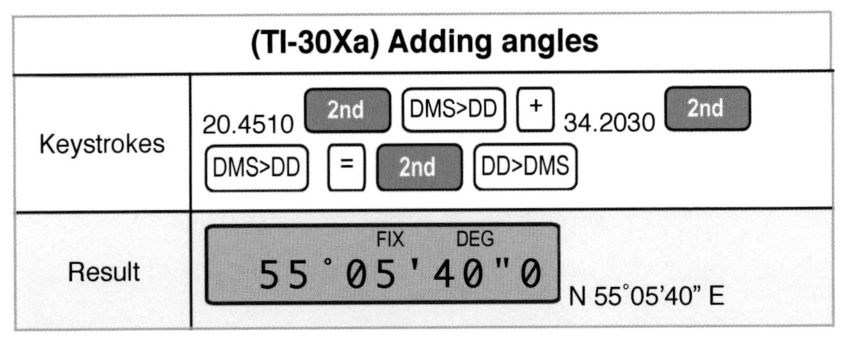

(TI-30Xa) Adding angles	
Keystrokes	20.4510 [2nd] [DMS>DD] [+] 34.2030 [2nd] [DMS>DD] [=] [2nd] [DD>DMS]
Result	FIX DEG 55°05'40"0 N 55°05'40" E

NOTE: When adding angles to bearings it is helpful to draw a sketch to aid in determining the quadrant for the final bearing.

See "Bearings and Azimuths - Book 1" for more information and equations on this topic.

Create Rectangular Coordinates

Given: S 51°50'34" E 175.00'
N1 = 10000, E1 = 10000
Find: N2 = ?, E2 = ?

Equation 2: N2 = (Cos(51°50'34") x 175.00') + 10000
Equation 3: E2 = (Sin(51°50'34") x 175.00') + 10000

(HP 35s) Courses	
Keystrokes	51.5034 [↰] [HMS->] [COS] 175.00 [X] [+/-] 10000 [+]
Result	RPN Y 0.00000 X 9,891.88124 N2
Keystrokes	51.5034 [↰] [HMS->] [SIN] 175.00 [X] 10000 [+]
Result	RPN Y 9,891.88124 X 10,137.60572 E2

(TI-30Xa) Courses	
Keystrokes	51.5034 [2nd] [DMS>DD] [COS] [X] 175.00 [=] [+⇄-] [+] 10000 [=]
Result	FIX DEG 9891.88124 N2

Keystrokes	51.5034 [2nd] [DMS>DD] [SIN] [x] 175.00 [=] [+] 10000 [=]
Result	FIX DEG 10137.60572 E2

NOTE: During the keystrokes for N2, the [+/-] (HP) and [+⇄-] (TI) was pressed before the N1 value (10000) was entered. This is because the course is in a Southeast direction therefore the Latitude is negative. This is the tricky part where most examinees forget to add algebraically correctly.

Inverse Between Rectangular Coordinates

Given: N1 = 11382.34876, E1 = 10000.38621
N2 = 10265.34872, E2 = 8712.37652
Find: Course between N1E1 and N2E2

Equation 4: Lat = 10265.34872 - 11382.34876
Equation 5: Dep = 8712.37652 - 10000.38621
Equation 6: Distance = $\sqrt{(Lat^2 + Dep^2)}$
Equation 7: Bearing = Arctan(Dep / Lat)

(HP 35s) Inverse	
Keystrokes	10265.34872 [Enter] 11382.34876 [-] [↱] [STO] [1]
Keystrokes	8712.37652 [Enter] 10000.38621 [-] [↱] [STO] [2]

Keystrokes	RCL 1 ↱ x^2 RCL 2 ↱ x^2 + $\sqrt{x}$
Result	RPN Y 0.00000 X 1,704.89239 Distance
Keystrokes	RCL 2 RCL 1 ÷ ↱ ATAN ↱ ->HMS
Result	RPN Y 1,704.89239 X 49.04019 S 49°04'01.9" W

(TI-30Xa) Inverse	
Keystrokes	10265.34872 - 11382.34876 = STO 1
Keystrokes	8712.37652 - 10000.38621 = STO 2
Keystrokes	RCL 1 x^2 + RCL 2 x^2 = $\sqrt{x}$
Result	FIX DEG 1704.89239 Distance
Keystrokes	RCL 2 ÷ RCL 1 = 2nd TAN^{-1} 2nd DD>DMS
Result	FIX DEG 49°04'01"9 S 49°04'01.9" W

NOTE: The result for Equation 4 is negative therefore the bearing is South. The result for Equation 5 is negative therefore the bearing is West. The trick is to watch these numbers as you are calculating them so you know what quadrant you are in. You can always recall the stored values to confirm.

See "Create Rectangular Coordinates - Book 2" and "Inverse Between Rectangular Coordinates - Book 3" for more information and equations on this topic.

Circular Curves

Given: Δ = 10°31'20" RT, R = 2864.78898
Find: L = (Δ * R * π) / 180

Equation 8: L = (10°31'20" x 2864.78898 * π) / 180

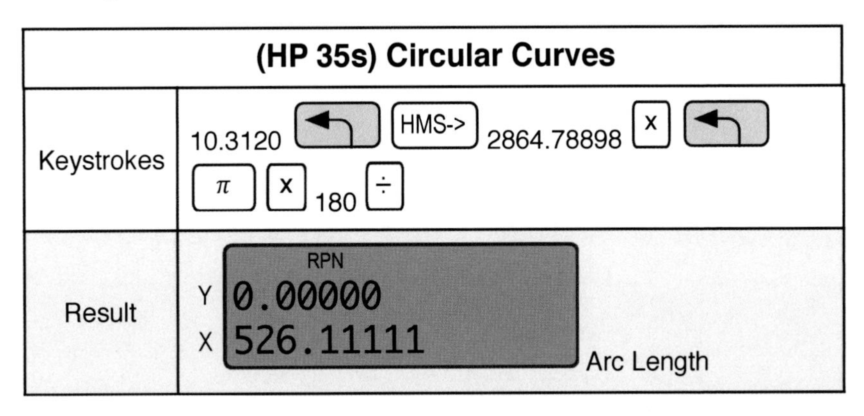

(HP 35s) Circular Curves	
Keystrokes	10.3120 [↰] [HMS->] 2864.78898 [x] [↰] [π] [x] 180 [÷]
Result	Y 0.00000 X 526.11111 Arc Length

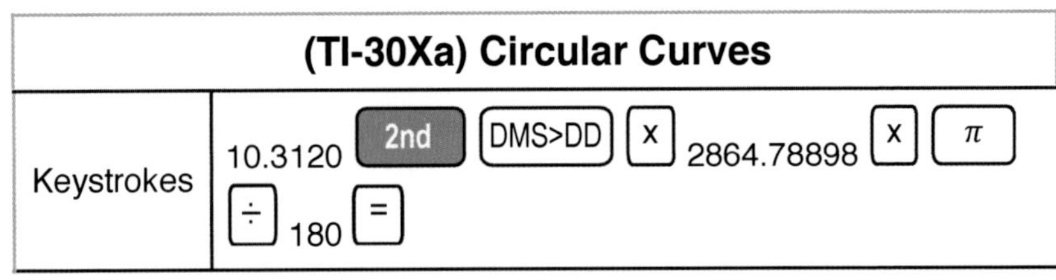

(TI-30Xa) Circular Curves	
Keystrokes	10.3120 [2nd] [DMS>DD] [x] 2864.78898 [x] [π] [÷] 180 [=]

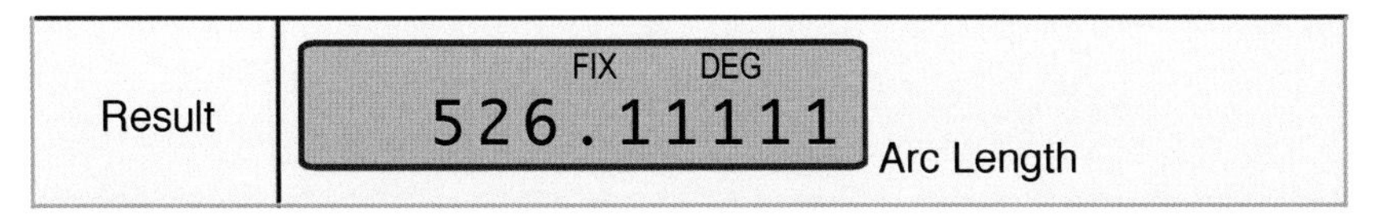

See "Circular Curves - Book 4" for more information and equations on this topic.

Spiral Curves

Given: Δt = 36°29'16", R = 2864.78898, O = 0.58160, T = 99.99594
Find: Ts = (Tan(Δt / 2) * (R + O)) + T

Equation 9: Ts = (Tan(36°29'16" / 2) x (2864.78898 + 0.58160)) + 99.99594

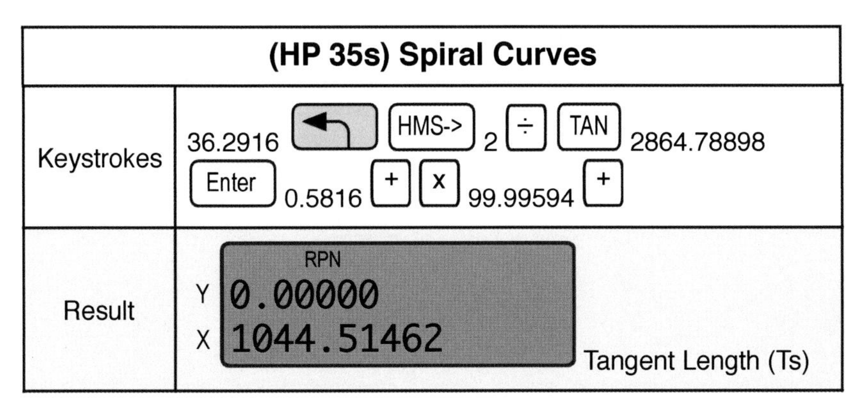

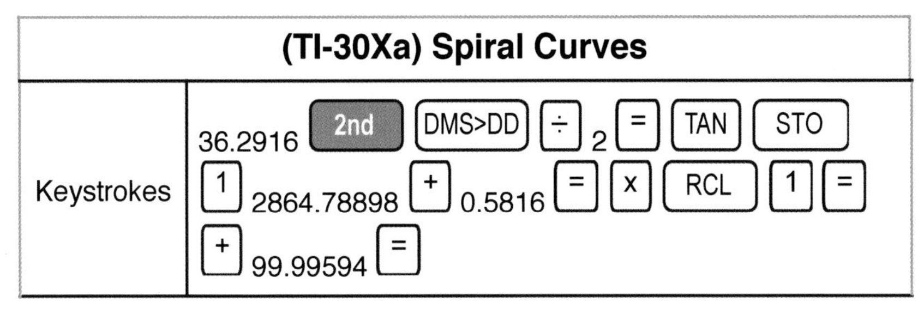

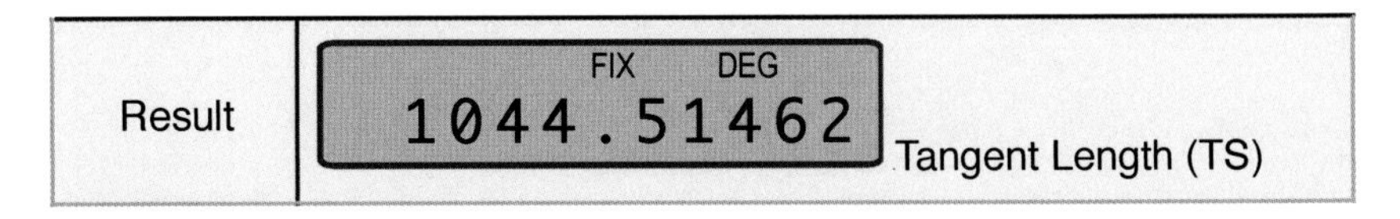

See "Spiral Curves - Book 6" for more information and equations on this topic.

Vertical Curves

Given: PVCElev = 108.00, G1 = -0.04, X = 266.67, R = 0.00015
Find: Y = PVCElev + G1(X) + (R/2)*X^2

Equation 10: Y = 108.00 + -0.04 x 266.67 + (0.00015 / 2) x 266.67^2

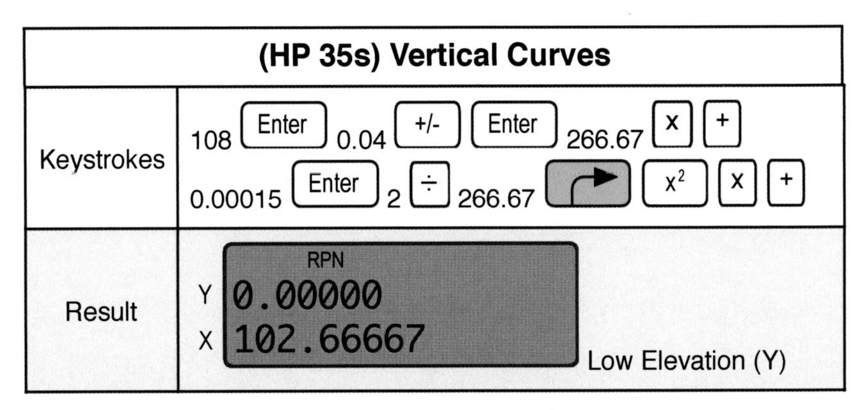

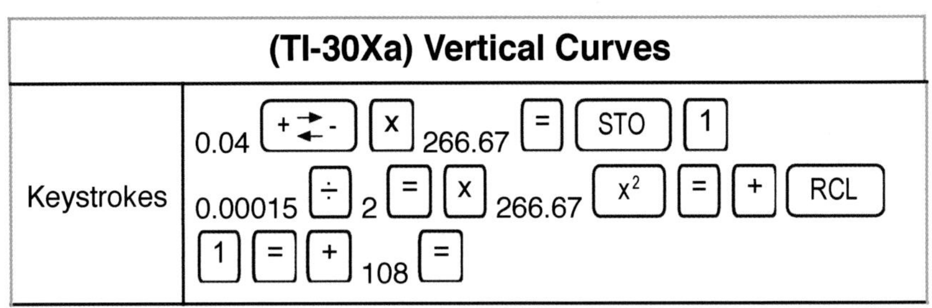

See "Vertical Curves - Book 10" for more information and equations on this topic.

Exercise 3

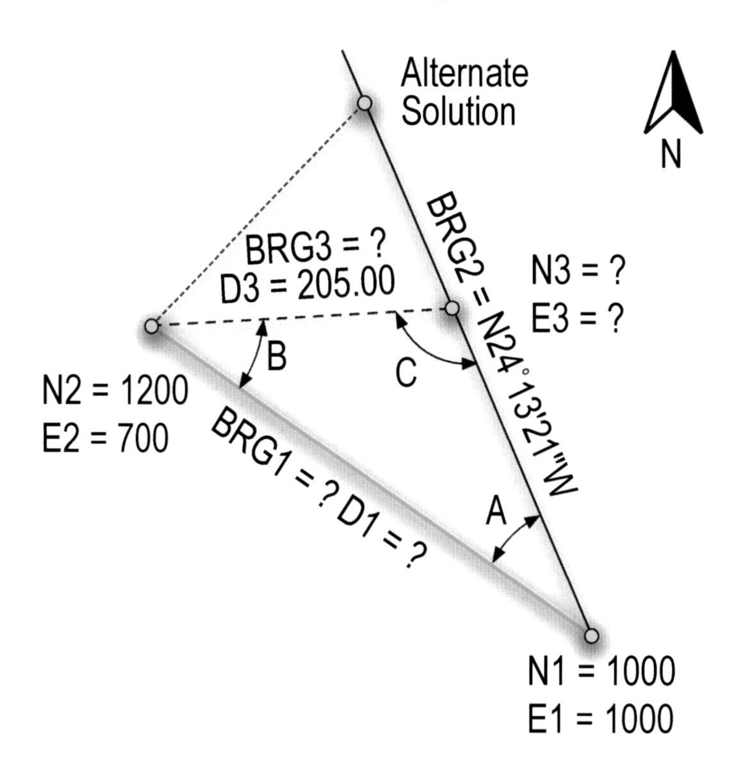

Figure 3.

Find: N3 = ?, E3 = ?

The answers can be found at the end of the book.

SOLUTIONS TO EXERCISES

Exercise 1 (DMS to D.ddd) Answers		
DMS	**HP 35s**	**TI-30Xa**
78°33'12"	RPN Y 0.00000 X 78.55333	FIX DEG 78.55333
123°12'56"	RPN Y 0.00000 X 123.21556	FIX DEG 123.21556
89°39'54"	RPN Y 0.00000 X 89.66500	FIX DEG 89.66500

Exercise 2 (D.ddd to DMS) Answers		
D.ddd	**HP 35s**	**TI-30Xa**
92.34521°	RPN Y 0.00000 X 92.20428 92°20'42.8"	FIX DEG 92°20'42"7 92°20'42.7"
98.89765°	RPN Y 0.00000 X 98.53515 98°53'51.5"	FIX DEG 98°53'51"5 98°53'51.5"
129.65754°	RPN Y 0.00000 X 129.39271 129°39'27.1"	FIX DEG 129°39'27" 129°39'27"

NOTE: Both calculators are set to 5 decimal places which affects the rounding for the tenths of a second differently.

Exercise 3 solution

Start by defining the equations for each step then solve each equation.

Given (See Figure 3):
N1 = 1000, E1 =1000, N2 = 1200, E2 = 700,
BRG2 = N 24°13'21" W, D3 = 205.00
Find: N3 = ?, E3 = ?

Equation 1: Lat = 1200 - 1000
Equation 2: Dep = 700 - 1000
Equation 3: $D1 = \sqrt{(Lat^2 + Dep^2)}$
Equation 4: BRG1= Arctan(Dep / Lat)

(HP 35s) Inverse - BRG1 & D1	
Keystrokes	1200 [Enter] 1000 [-] [↱] [STO] [1]
Keystrokes	700 [Enter] 1000 [-] [↱] [STO] [2]
Keystrokes	[RCL] [1] [↱] [x^2] [RCL] [2] [↱] [x^2] [+] [$\sqrt{x}$]
Result	RPN Y 0.00000 X 360.55513 Distance (D1)
Keystrokes	[RCL] [2] [RCL] [1] [÷] [↱] [ATAN] [↱] [->HMS]

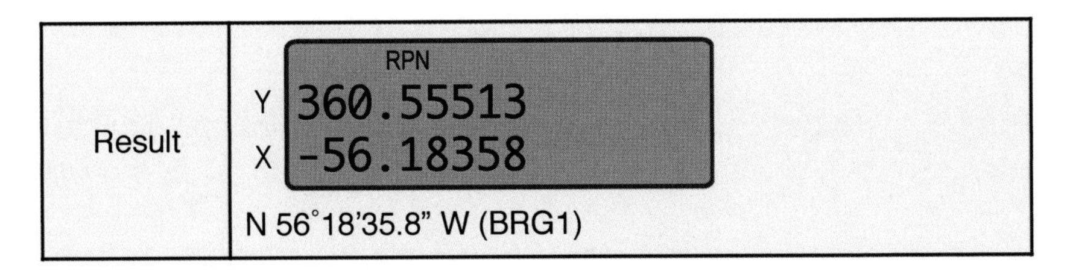

Result	RPN Y 360.55513 X -56.18358 N 56°18'35.8" W (BRG1)

(TI-30Xa) Inverse - BGR1 & D1	
Keystrokes	1200 [-] 1000 [=] [STO] [1]
Keystrokes	700 [-] 1000 [=] [STO] [2]
Keystrokes	[RCL] [1] [x^2] [+] [RCL] [2] [x^2] [=] [$\sqrt{x}$]
Result	FIX DEG 360.55513 Distance (D1)
Keystrokes	[RCL] [2] [÷] [RCL] [1] [=] [2nd] [TAN^{-1}] [2nd] [DD>DMS]
Result	FIX DEG -56°18'35"7 N 56°18'35.7" W (BRG1)

Find: A = BRG1 - BRG2

Equation 5: A = N 56°18'35.7" W - N 24°13'21" W

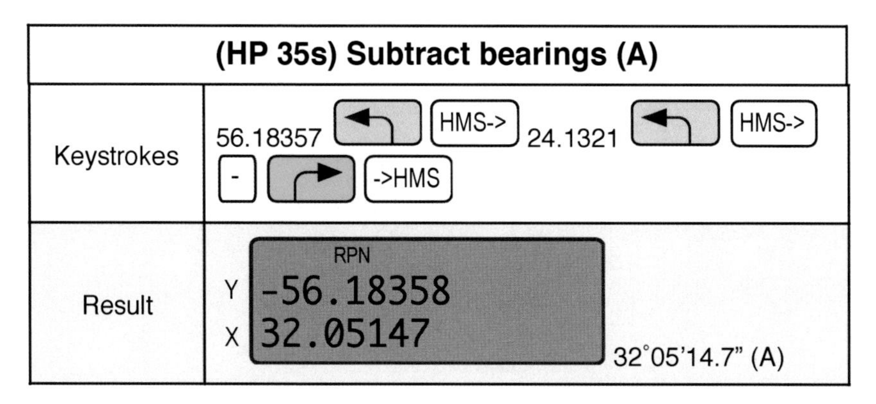

(HP 35s) Subtract bearings (A)	
Keystrokes	56.18357 [↰] [HMS->] 24.1321 [↰] [HMS->] [-] [↱] [->HMS]
Result	RPN Y -56.18358 X 32.05147 32°05'14.7" (A)

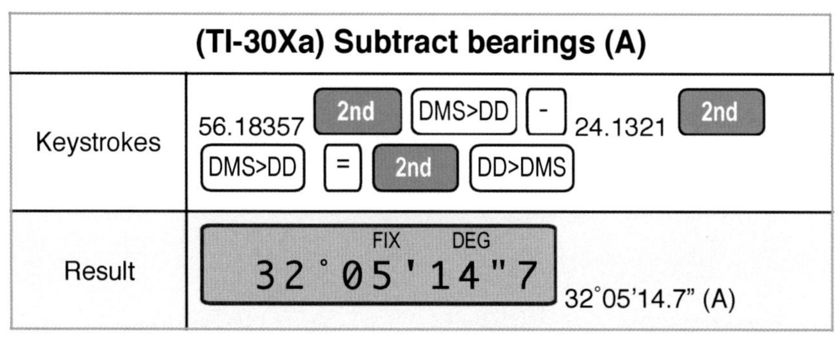

(TI-30Xa) Subtract bearings (A)	
Keystrokes	56.18357 [2nd] [DMS>DD] [-] 24.1321 [2nd] [DMS>DD] [=] [2nd] [DD>DMS]
Result	FIX DEG 32°05'14"7 32°05'14.7" (A)

Find: C = ArcSin(SIN(A) x D1 / D3)

Equation 6: C = ArcSin(SIN(32°05'14.7") x 360.55513 / 205)

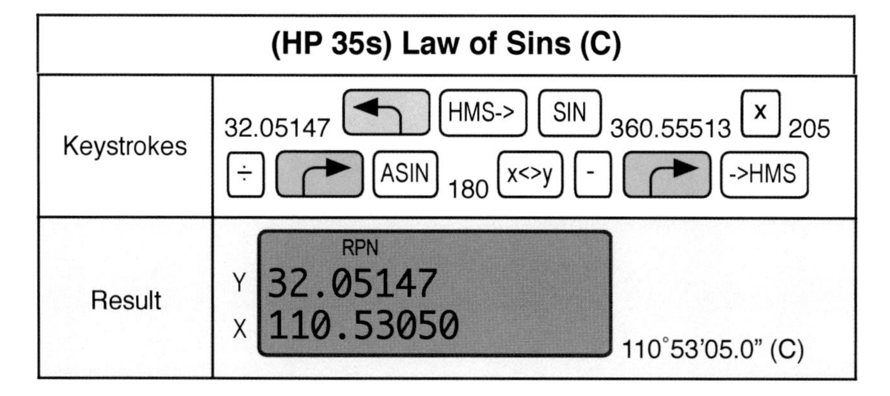

(HP 35s) Law of Sins (C)	
Keystrokes	32.05147 [↰] [HMS->] [SIN] 360.55513 [x] 205 [÷] [↱] [ASIN] 180 [x<>y] [-] [↱] [->HMS]
Result	RPN Y 32.05147 X 110.53050 110°53'05.0" (C)

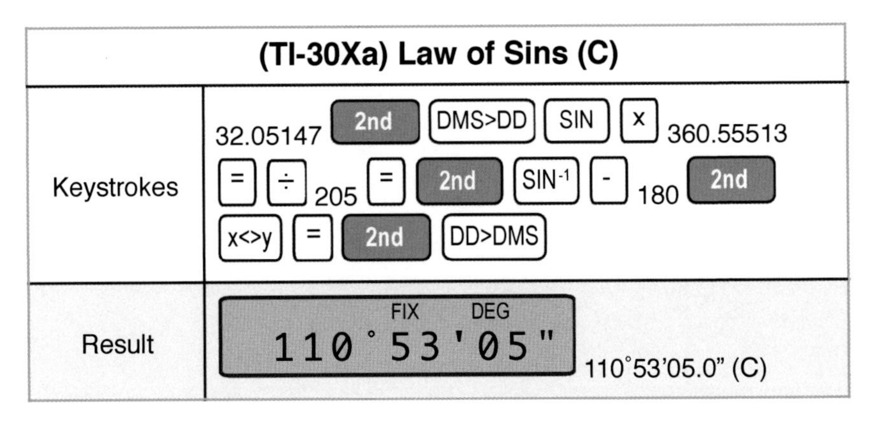

(TI-30Xa) Law of Sins (C)	
Keystrokes	32.05147 [2nd] [DMS>DD] [SIN] [x] 360.55513 [=] [÷] 205 [=] [2nd] [SIN-1] [-] 180 [2nd] [x<>y] [=] [2nd] [DD>DMS]
Result	FIX DEG 110°53'05" — 110°53'05.0" (C)

NOTE: In Figure 3 the angle for C is greater than 90° therefore the angle that is initially generated needs to be subtracted from 180°. When using the Law of Sins you need to be aware that there are two solutions for a Bearing-Distance Intersection.

Find: B = 180 - A - C

Equation 7: B = 180 - 32°05'14.7" - 110°53'05.0"

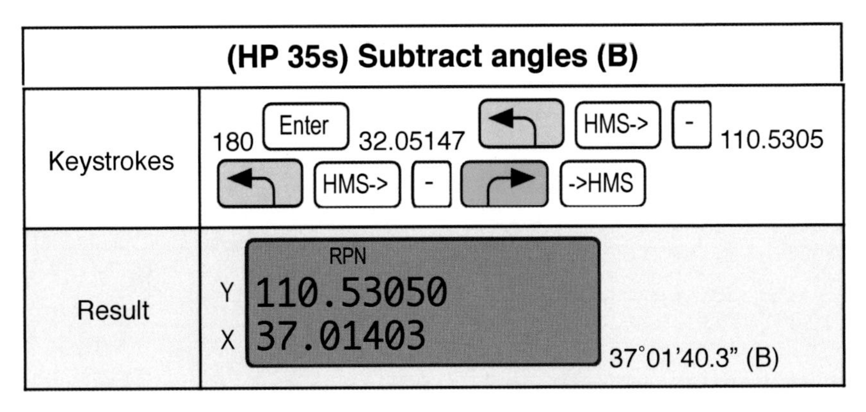

(HP 35s) Subtract angles (B)	
Keystrokes	180 [Enter] 32.05147 [↰] [HMS->] [-] 110.5305 [↰] [HMS->] [-] [↱] [->HMS]
Result	RPN Y 110.53050 X 37.01403 — 37°01'40.3" (B)

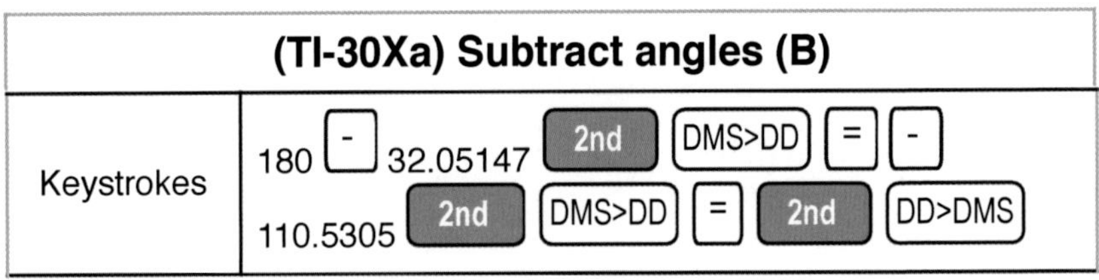

(TI-30Xa) Subtract angles (B)	
Keystrokes	180 [-] 32.05147 [2nd] [DMS>DD] [=] [-] 110.5305 [2nd] [DMS>DD] [=] [2nd] [DD>DMS]

Result	FIX DEG 37°01'40"3 37°01'40.3" (B)

Find: BRG3 = BRG1 + B

Equation 8: BRG3 = S 56°18'35.8" E + 37°01'40.3"

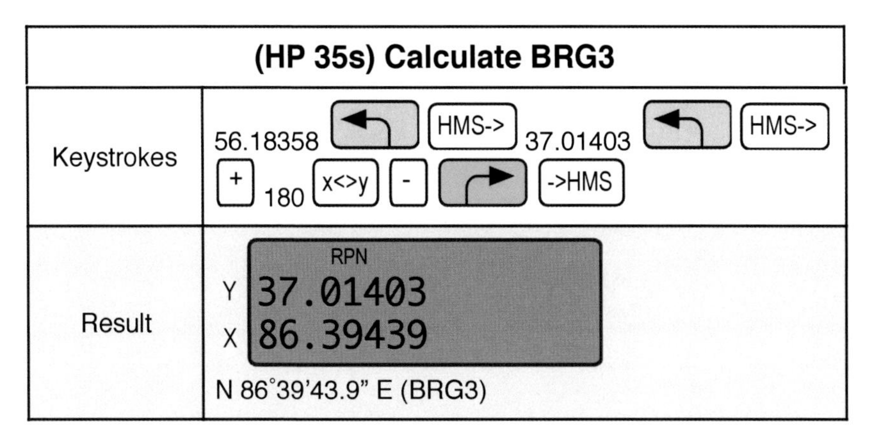

(HP 35s) Calculate BRG3	
Keystrokes	56.18358 ↰ HMS-> 37.01403 ↰ HMS-> + 180 x<>y - ↱ ->HMS
Result	RPN Y 37.01403 X 86.39439 N 86°39'43.9" E (BRG3)

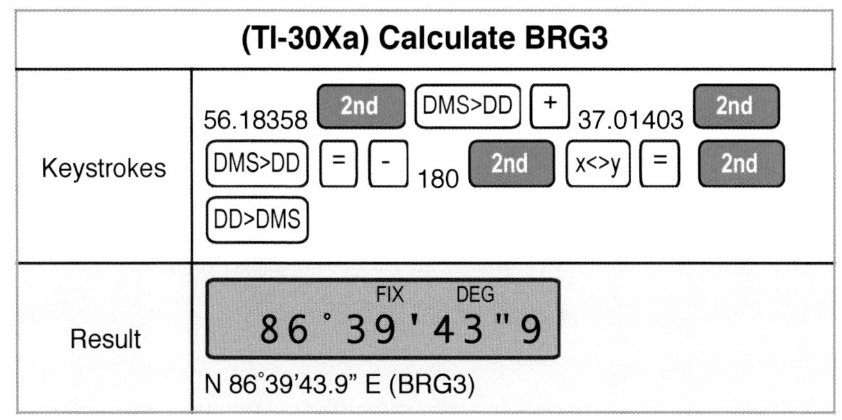

(TI-30Xa) Calculate BRG3	
Keystrokes	56.18358 2nd DMS>DD + 37.01403 2nd DMS>DD = - 180 2nd x<>y = 2nd DD>DMS
Result	FIX DEG 86°39'43"9 N 86°39'43.9" E (BRG3)

NOTE: The addition goes over 90° therefore we have to subtract from 180° to get the bearing in the correct quadrant.

Find: N3 = ?, E3 = ?

Equation 9: N3 = (Cos(86°39'43.9") x 205.00') + 1200

Equation 10: E3 = (Sin(86°39'43.9") x 205.00') + 700

(HP 35s) Calculate N3 & E3	
Keystrokes	86.39439 [↰] [HMS->] [COS] 205 [x] 1200 [+]
Result	RPN Y 0.00000 X 1,211.93566 N3
Keystrokes	86.39439 [↰] [HMS->] [SIN] 205.00 [x] 700 [+]
Result	RPN Y 0.00000 X 904.65224 E3

(TI-30Xa) Calculate N3 & E3	
Keystrokes	86.39439 [2nd] [DMS>DD] [COS] [x] 205.00 [=] [+] 1200 [=]
Result	FIX DEG 1211.93566 N3
Keystrokes	86.39439 [2nd] [DMS>DD] [SIN] [x] 205.00 [=] [+] 700 [=]
Result	FIX DEG 904.65224 E3

See "Bearings and Azimuths - Book 1", "Create Rectangular Coordinates - Book 2", "Inverse Between Rectangular Coordinates - Book 3", "Intersections - Book 8" and "What was that formula? - Book 11" for more information and equations on this topic.

CONCLUSION

The intent of this book is to show that performing manual calculations is not that difficult once you master the steps.

This book would be quite lengthy to show the calculator steps for all of the surveying equations that you will use during the course of your career. We have demonstrated the calculator entries for general surveying equations. Over time you will use more calculator functions to speed up the process such as Polar to Rectangular conversions which is not on every calculator. We have only covered the bare minimum needed to become a Master of the Calculator for the exam.

The Key notes are as follows:

- Select a calculator that has DMS <> D.ddd conversion function.
- Convert DMS to D.ddd before performing trigonometric functions.
- Convert D.ddd to DMS for the final answer.
- Remember to add algebraically.
- Learn the math.
- Draw sketches to determine which quadrant you are working in.
- Most of all Practice, Practice, Practice

The calculator steps will become automatic the more you perform the calculations. There is no substitute for practice.

There are no shortcuts or calculator programs that will replace the knowledge that is needed to pass the exam.

Be the best surveyor you can be and earn respect from your peers. Learn from the experts, learn the math, master the calculator and most of all have fun.

Happy Calculating...

ABOUT THE AUTHOR

Jim Crume P.L.S., M.S., CFedS

My land surveying career began several decades ago while attending Albuquerque Technical Vocational Institute in New Mexico and has traversed many states such as Alaska, Arizona, Utah and Wyoming. I am a Professional Land Surveyor in Arizona, Utah and Wyoming. I am an appointed United States Mineral Surveyor and a Bureau of Land Management (BLM) Certified Federal Surveyor. I have many years of computer programming experience related to surveying.

This ebook is dedicated to the many individuals that have helped shape my career. Especially my wife Cindy. She has been my biggest supporter. She has been my instrument person, accountant, advisor and my best friend. Without her, I would not be the professional I am today. Cindy, thank you very much.

Other titles by this author:

http://www.cc4w.net/ebooks.html

Follow us on Facebook

Books available on amazon.com

SURVEYING MATHEMATICS MADE SIMPLE

MATH-SERIES TRAINING AND REFERENCE BOOKS / APPS

Printed - Digital - Apps

Many Titles to choose from.

www.cc4w.net

A New Math-Series of books with useful formulas, helpful hints and easy to follow step by step instructions.

www.facebook.com/surveyingmathematics

Digital and Printed Editions Math-Series Training and Reference Books. Designed and written by Surveyors for Surveyors, Land Surveyors in Training, Engineers, Engineers in Training and aspiring Students.